Cristina Ivaldi

# BugabooO

## The Wicked Witch

U0061908

 商務印書館

# Contents

Originally published by Black Cat Publishing under the title: *Bugaboo The Wicked Witch*
© 2001 Black Cat Publishing
An imprint of Cideb Editrice, Genoa, Canterbury

Name of Book: Bugaboo The Wicked Witch
Author: Cristina Ivaldi
Editors: Claudia Fiocco, Kathryn Pass
Design and Art Direction: Nadia Maestri
Illustrations: Claudio Decataldo
Layout: Sara Blasigh

系 列 名： Quality English Learning for Kids · I
書　　 名： Bugaboo The Wicked Witch
責任編輯： 傅　伊
出　　 版： 商務印書館 (香港) 有限公司
　　　　　 香港筲箕灣耀興道 3 號東滙廣場 8 樓
　　　　　 http://www.commercialpress.com.hk
印　　 刷： 美雅印刷製本有限公司
　　　　　 九龍觀塘榮業街 6 號海濱工業大廈 4 樓 A
版　　 次： 2004 年 9 月第 1 版第 1 次印刷
　　　　　 © 2004 商務印書館 (香港) 有限公司
　　　　　 ISBN 962 07 1718 X
　　　　　 Printed in Hong Kong

# 出版説明

　　學英語當然要學優質的，有品質才能讓人有信心。我們一直積極提倡學習優質英語的理念，並且為學習者提供過多元化的優質英語材料，像《Black Cat 優質英語階梯閱讀》就十分成功，至今已出版近60本。鑑於良好的英語能力最好從小培養，我們於是出版這一套適合五至八歲兒童的優質英語閱讀讀本 "Quality English Learning for Kids・I"。

　　培養兒童對於英語的興趣，須從趣味和簡易兩方面入手。圖文並茂，聲文結合這兩大特點對學習英語甚有幫助。"Quality English Learning for Kids・I" 承續本館出版優質英語書的理念，全書彩圖精美，附 CD 朗讀內容及聆聽練習，形式多元化，有出版故事讀本（story books）、圖畫讀本（picture readers）、戲劇讀本（drama readers）及互動讀物（interactive readers）四大類，提供不同學習功能。故事讀本和圖畫讀本可供兒童看圖講故事；戲劇讀本完全用對白編寫，培養脫口而出講英語的習慣，適合家庭裏作簡單的角色扮演，或者小學生在課堂上作簡單的演出。

　　針對兒童學習英語的需要，本系列提示家長為兒童設定學習目標，並且説明如何達標，另備生詞表和語法知識點，讓兒童在家長協助下掌握生詞用法，認識簡單的句子結構和了解語法要點。

　　"Quality English Learning for Kids・I" 吸引兒童對閱讀產生興趣，逐步引導他們參與愉快的閱讀旅程。在這個旅程中，家長是重要的導航者，透過對兒童的悉心鼓勵，循循善誘，進一步加強親子關係。

商務印書館
編輯部

# 使用説明

① **如何使用本書?**

本書為圖畫讀本(picture reader),適合課堂使用或親子共讀。

每頁均圖文並茂。正文包括對話和敘述文字,讓小孩子熟悉並掌握第一、二、三身句法。老師或家長可讓小孩子用第一身敘述句法作自我介紹,用第二身疑問句法提問,用第三身敘述句法介紹親人、朋友,或描述某個物體。老師或家長還可與小孩子進行簡單對話,練習使用第一、二、三身句法。

除正文外,還設有看圖練習題(辨識、填空、配對等),培養小孩子的觀察、判斷和記憶能力。老師或家長可指出實物,讓小孩子說出它們的名稱。

本書配有 CD,小孩子可邊聽邊讀,提高英語聽説能力。

② **本書的學習目標是甚麼?**

老師或家長可為孩子定出以下學習目標。

使用本書後,孩子學會:

(a) 說出遊樂場常見設施的名稱(say the items of funfair);

(b) 辨認不同地方(recognize different places);

(c) 說出某人住在哪裏(say where people live);

(d) 聽從CD的指示,找出正確的圖片或路線(The Funfair & Lost in the Funfair)和玩填空(Where do they live?)、配對(The Broken Signs)等遊戲。

③ **本書有哪些重點生詞和語法學習項?**

(a) 重點生詞:本書的重點生詞包括四大類,即遊樂場設施(funfair facilities)、英語字母(the alphabet)、地理位置(places)和描述事物情狀的形容詞(adjectives)。另附英漢對照生詞表,幫助理解和記憶生詞。

(b) 語法學習項:

第一身句法(the first person)(例如頁3,"I'm Bugaboo the witch!")

第二身句法(the second person)(例如頁5,"Are you hurt?")

第三身句法(the third person)(例如頁3,"This is Bugaboo. She is a witch.")

簡單現在時(present simple)動詞為原形,但當主語為第三身單數名詞或代詞時,動詞為has、is等特殊形式或在詞尾加s。(例如頁6,"She wants to help." 和頁8,"She has powerful spells!")

注: 故事完結後,CD Track 10-22 朗讀生詞,例如:

Track 10  Page 3, Vocabulary, chairlift

Track 11  Page 6, Vocabulary, snow, ski

This is Bugaboo.
She is a witch.
She is a bad witch, very bad!
She is a wicked witch!

Bugaboo is wicked,
but she is not very <u>clever</u>.
She is not clever at all!

AH AH–
ARGH!

Now Bugaboo is a **friendly** witch,
not a **wicked** witch!
She wants to help.

The <u>snow</u>  is <u>cold</u>! 🗻 <u>Skiers</u> 🎿 are cold!

SNOW

SKI

MY FRIEND!
CAN I HELP?

The snow is cold.
The skier is cold.
But Bugaboo is a <u>powerful</u> witch.

She has powerful <u>spells</u>!

1. SNOW GO AWAY! A-W-A-Y

SPELL

2. BOOOOO! BOOOOO!

There are <u>a</u> <u>lot</u> <u>of</u> <u>things</u>  to do.

There is a <u>duckling</u> on the <u>river</u> <u>b</u>ank.

And <u>mother</u> <u>duck</u> is on the other bank.

Poor duckling. All that water!

1. WATER GO AWAY! A-W-A-Y

2. OUCH! WHERE'S THE WATER?

Bugaboo is very happy.
She is a friendly witch
and she is famous, too!

Oh! A <u>fire</u>!

The fire is <u>dangerous</u>.
No problem!
Bugaboo is ready!

The seaside!
And what is this?
A <u>fish</u>  in the <u>net</u>?
Poor fish!
Bugaboo can help!
Or is it a <u>shark</u>?

**1.** POOR LITTLE FISH!

**2.** IT'S A SHARK!

**3.** IT'S A SHARK! WE'VE GOT IT!

**4.** LOOK!

BOAT

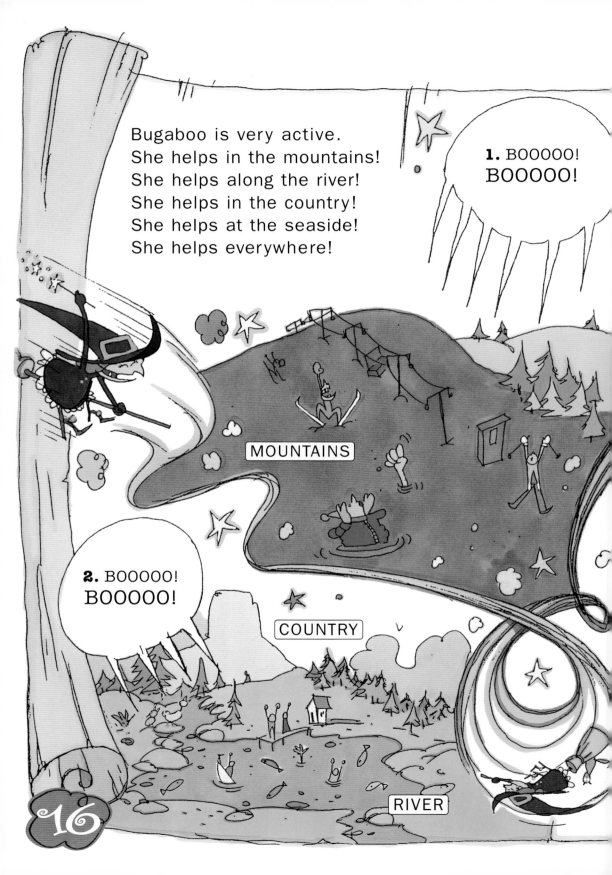

She helps in <u>town</u> too!
Oh! The <u>traffic</u> <u>lights</u>!
They are <u>red</u>!
People are <u>bored</u>!
People are <u>tired</u>!
But Bugaboo can help!
Now all the lights are <u>green</u>.

But what is that?
Is it a <u>dragon</u>?
No, it is not a dragon!
It is the <u>roller</u> <u>coaster</u>!
This is a funfair! But Bugaboo wants to help!
There is a <u>gate</u>.
And a big <u>sign</u>.
And Bugaboo is very <u>fast</u>!

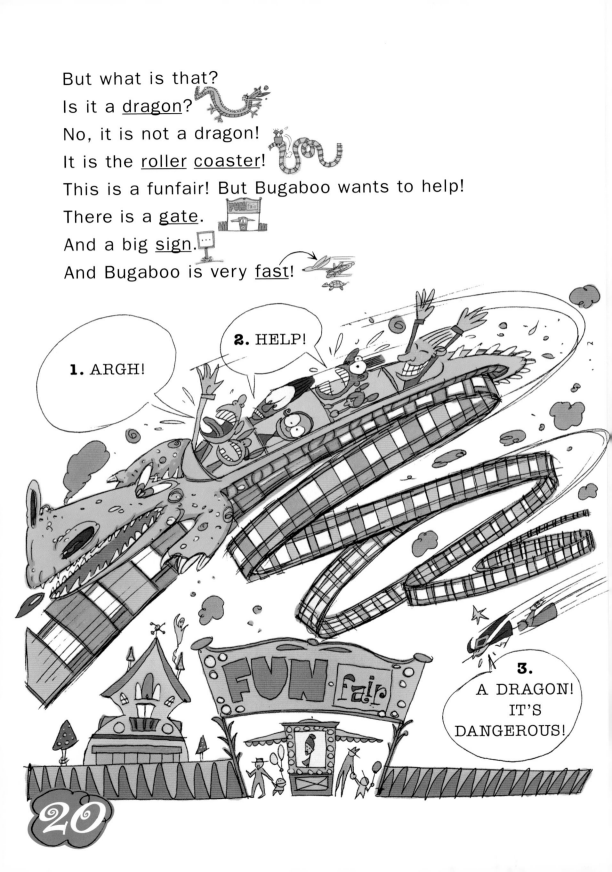

THUD!

21

Bugaboo is wicked again!
She is a real witch. A wicked witch.
And now people are <u>happy</u> again.
Bugaboo is wicked, but she is <u>stupid</u>.
A stupid witch is not dangerous!

# The Funfair

Listen to each word, then write it in the space next to the correct picture.

**big wheel**

**chips**

**bumper cars**

**hot dog**

**witch's house**

stall

candy floss

roundabout

ghost train

roller coaster

25

# Lost in the Funfair

Help the children find their mums.
Listen to the recording and follow the letters.

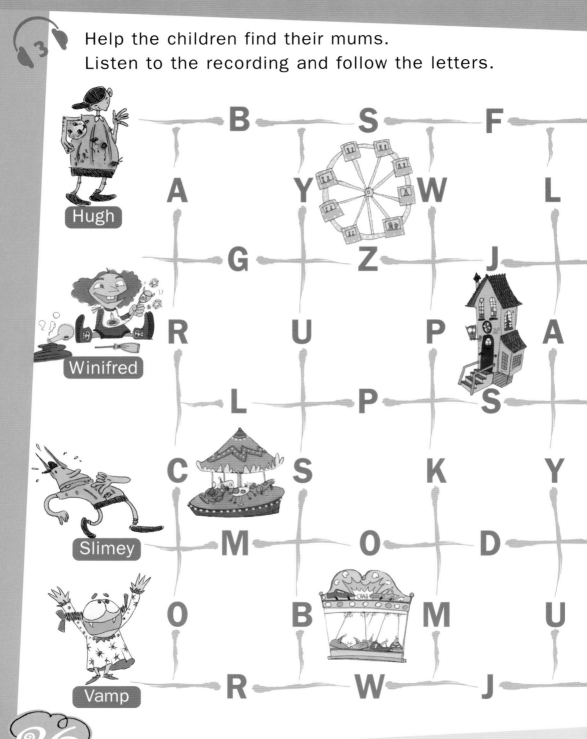

Hugh

B S F

A Y W L

G Z J

Winifred

R U P A

L P S

Slimey

C S K Y

M O D

Vamp

O B M U

R W J

# QUALITY ENGLISH CLUB

## Membership Application Form

**QUALITY ENGLISH CLUB** is for those who love English reading and seek for better English to share and learn with fun together.

**Benefits offered:**
- Membership Card
- English learning e-forum
- English learning activities
- Surprise gift and more...

Simply fill out the application form below and fax it back to 2565 1113 or send it back to the address at the back.

**Join Now! It's FREE** exclusively for readers who have purchased books on Quality English Learning published by the Commercial Press!

(Please fill out the form with **BLOCK LETTERS**.)

The title of book(s) /book set(s) that you have purchased: _____

English Name: _____ (Surname) _____ (Given Name)

Chinese Name: _____

Address:

Tel: _____ Fax: _____

Email: _____

Sex: ❏ Male ❏ Female    (Login password for e-forum will be sent to this email address.)

Education Background: ❏ Kindergarten   ❏ Primary 1-3   ❏ Primary 4-6
❏ Junior Secondary Education (F1-3)   ❏ Senior Secondary Education (F4-5)
❏ Matriculation   ❏ College   ❏ University or above

Age: ❏ 3 - 6   ❏ 6 - 9   ❏ 10 - 12   ❏ 13 - 15   ❏ 16 - 18
❏ 19 - 24   ❏ 25 - 34   ❏ 35 - 44   ❏ 45 - 54   ❏ 55 or above

Occupation: ❏ Student   ❏ Teacher   ❏ White Collar   ❏ Blue Collar
❏ Professional   ❏ Manager   ❏ Business Owner   ❏ Housewife
❏ Others (please specify: _____ )

As a member, what would you like **QUALITY ENGLISH CLUB** to offer:
❏ Member gathering/ party   ❏ English class with native teacher   ❏ English competition
❏ Newsletter   ❏ Online sharing   ❏ Book fair
❏ Book discount   ❏ Others (please specify: _____ )

Other suggestions to **QUALITY ENGLISH CLUB**: _____

_____

Please sign here: _____ (Date: _____ )

Visit us at Quality English Learning Online http://publish.commercialpress.com.hk/qel

# QUALITY ENGLISH CLUB

The Commercial Press (Hong Kong) Ltd.
8/F, Eastern Central Plaza,
3 Yiu Hing Road, Shau Kei Wan,
Hong Kong

THE COMMERCIAL PRESS (H.K.) LTD.

27

# The Labyrinth

Throw the dice 🎲 🎲 and follow the instructions.

START

COUNTRY

TREES

MOUNTAINS

GO TO THE FUNFAIR

ROLLER COASTER

GO TO THE COUNTRY

SING A SONG

FINISH

RIVER

MISS 1 TURN

GO TO THE MOUNTAINS

TRAFFIC LIGHTS

29

# Where do they live?

Who lives in the country?
Who lives in town?
Who lives at the funfair?
Who lives at the seaside?
Who lives in the mountains?

Listen to the recording and complete the sentences with their names.

**6** Morgan the Witch

**5** Willy Wizard

**9** Queenie Witch

**8** Walter Wizard

**7** Wanda the Witch

..................................... lives
at the seaside.

..................................... lives
in town.

..................................... lives
at the funfair.

..................................... lives
in the country.

..................................... lives
in the mountains.

# The Broken Signs

Bugaboo has BRO KEN 8 signs with her head. What are they?

TRAIN

TRAFFIC

CANDY

COASTER

GHOST

CARS

WHEEL

WITCH'S

LIGHTS

ROLLER

DOG

BIG

FLOSS

BUMPER

HOUSE

HOT

# Bugaboo

## The Wicked Witch

## 🎧 The Funfair

*pages 24 and 25*

Listen to each word, then write it in the space next to the correct picture.

big wheel

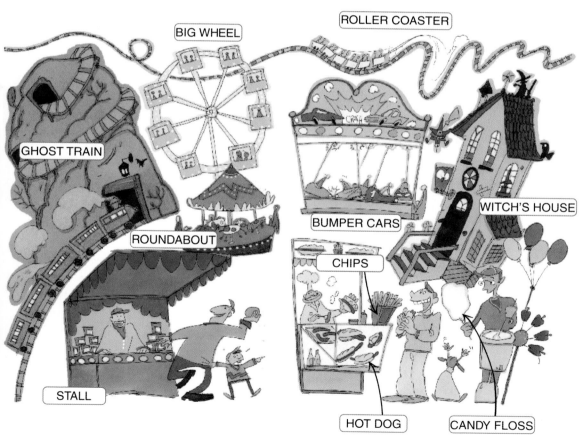

BIG WHEEL

ROLLER COASTER

GHOST TRAIN

WITCH'S HOUSE

BUMPER CARS

ROUNDABOUT

CHIPS

STALL

HOT DOG

CANDY FLOSS

chips
bumper cars
hot dog
witch's house
stall
candy floss
roundabout
ghost train
roller coaster

## 🎧 Lost in the Funfair

*pages 26 and 27*

Help the children find their mums.
Listen to the recording and follow
the letters.

HUGH:
B Y Z P S Y T G B J D

WINIFRED:
G U P K M J U T G A

SLIMEY:
M O K P W F L N X C Q

VAMP:
R B O M J L S H M B C E

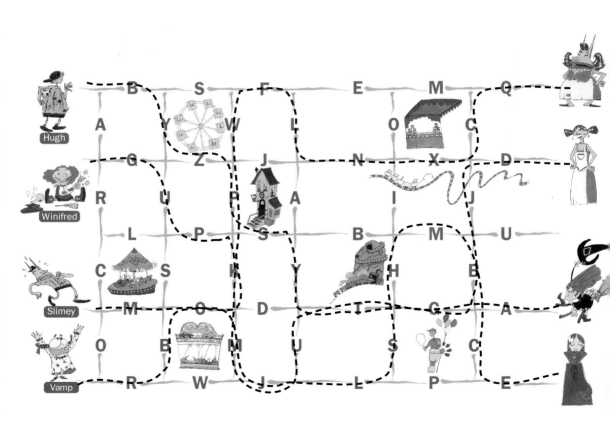

# The Labyrinth

*pages 28 and 29*

Throw the dice and follow the instructions.

**START**
country
trees
count to ten (10)
river and duck
town
go to town
go back 2 (two)
snow
go to the river
go to the mountains
miss 1 (one) turn
traffic lights
sing a song
go to the country
mountains
go to the funfair
roller coaster
funfair
go back 2 (two)
miss 1 (one) turn
car
river
**FINISH**

## 🎧 ⁴⁻⁹ Where do they live?

*pages 30 and 31*

Who lives in the country?
Who lives in town?
Who lives at the funfair?
Who lives at the seaside?
Who lives in the mountains?

Listen to the recording and complete the sentences with their names.

Morgan the Witch
Willy Wizard
Queenie Witch
Walter Wizard
Wanda the Witch

**My name's Willy**. Willy the Wizard. I live in the mountains. There's a lot of snow in winter and it's cold, in the mountains. But I like skiing and I like the snow.

**I'm Morgan the Witch**. I live at the seaside. There's a big beach and a lot of people come to swim in the sea. There are lots of fish in the sea, and I like fishing. I'm afraid of sharks, but there are no sharks here.

**My name's Wanda** and I'm a country witch. I live in the country. There is a river with lots of ducks, and fish too. There are green fields and trees and animals in the country. I think the country is beautiful.

**I'm Walter and I'm a Wizard.** I live in town. There are cars and traffic and traffic lights, but I like the town. I like the cinemas and the shops and the lights and the people. There are lots of people everywhere, day and night, in town.

**My name's Queenie** and I live at the funfair. Do you know the witch's house? That's where I live, in front of the big wheel. There's also a ghost train at the funfair. I like it a lot. But I don't like the roller coaster. I think it's dangerous.

35

- Morgan the Witch lives at the seaside.
- Walter Wizard lives in town.
- Queenie Witch lives at the funfair.
- Wanda the Witch lives in the country.
- Willy Wizard lives in the mountains.

# The Broken Signs

*page 32*

Bugaboo has BROKEN 8 signs with her head.
What are they?

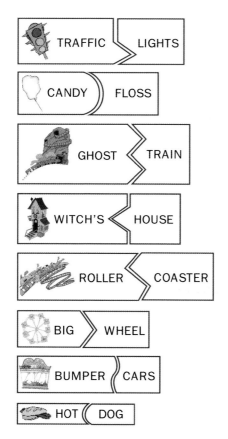

# GLOSSARY

active 積極的

a lot of 許多

bank 岸

bored 厭倦的

bumper cars 碰碰車

candy floss 棉花糖

chairlift （送滑雪者上、下山的）
升降椅

clever 聰明的

country 鄉村

dangerous 危險的

different 不同的

dragon 龍

duckling 小鴨子

famous 著名的

friendly 友好的

funfair 公共露天遊樂場

hot dog 熱狗

hungry 飢餓的

hurt 受傷的

labyrinth 迷宮

lost 迷路的

mountains 山區

no problem 沒關係

powerful 強有力的

roundabout 繞道

rucksack （旅行用的）帆布背包

seaside 海邊

ski 滑雪

spells 魔咒

stall 貨攤

stupid 愚蠢的

tent 帳篷

tired 疲勞的

town 城鎮

traffic lights 交通燈，又稱紅綠燈

train 火車

wheel 輪子

wicked 邪惡的

witch 女巫